The Scientific Revolution

LANCASTER PAMPHLETS

The Scientific Revolution

P. M. Harman

METHUEN · LONDON AND NEW YORK

First published in 1983 by
Methuen & Co. Ltd
11 New Fetter Lane,
London EC4P 4EE

Published in the USA by
Methuen & Co.
in association with Methuen, Inc.
733 Third Avenue, New York,
NY 10017

Typeset in Great Britain by
Scarborough Typesetting Services
and printed by
Richard Clay (The Chaucer Press)
Bungay, Suffolk

British Library Cataloguing in
Publication Data

Harman, P. M.
The scientific revolution.
– (Lancaster pamphlets)
1. Science – History
I. Title   II. Series
509   Q125

ISBN 0–416–35040–2

Contents

Foreword

Lancaster Pamphlets offer concise and up-to-date accounts of major historical topics, primarily for the help of students preparing for Advanced Level examinations, though they should also be of value to those pursuing introductory courses in universities and other institutions of higher education. They do not rely upon prior textbook knowledge. Without being all-embracing, their aims are to bring some of the central themes or problems confronting students and teachers into sharper focus than the textbook writer can hope to do; to provide the reader with some of the results of recent research which the textbook may not embody; and to stimulate thought about the whole interpretation of the topic under discussion. They are written by experienced university scholars who have a strong interest in teaching.

The Scientific Revolution

Introduction

The Scientific Revolution is the term traditionally used to describe the spectacular intellectual triumphs of sixteenth- and seventeenth-century European astronomy and physical science. By around 1700 educated men conceived the universe as a mechanical structure like a clock, the earth was regarded as a planet revolving round the sun, and the mysteries of nature were supposed to be open to investigation by means of experimentation and mathematical analysis. These new attitudes to the natural world contrast strikingly with the traditional conception of nature: that the earth was immobile and the centre of the cosmos, the cosmos itself being envisaged as a structure of crystalline spheres enveloping the central earth like the layers of an onion; nature was conceived as a living organism, a connected structure linked by a web of hidden active powers.

The major shift in outlook which led to emergence of the concept of the clockwork universe was closely associated with a broader cultural transformation, in which the acquisition of natural knowledge and the attainment of the control of nature were associated with man's religious destiny. The emergence of natural science in the early modern period as a seminal feature of European culture must be interpreted in the relation to the social and intellectual ferment of the period. Described by Herbert Butterfield as the most important event in European history since the rise of Christianity, the Scientific Revolution was as much a

1

cultural phenomenon as a revolution in scientific method and cosmology. The intellectual transformation associated with the Scientific Revolution led to a new confidence in the value of the investigation of nature and its control, a development which is fundamental to an understanding of the importance of science in modern society.

The seventeenth century was characterized by an unprecedented optimism about the potential for human advancement through technological improvement and an understanding of the natural world. Hopes were expressed that the understanding and control of nature would improve techniques in industry and agriculture. There was however a large gap between intention and achievement in the application of scientific knowledge. While assertions of the practical usefulness of natural knowledge and its prospective implications for technological improvement were common, the cultivation of science had little effect on the relationship between man and his environment in the seventeenth century. Nevertheless the cultural values associated with the pursuit of natural knowledge were a significant characteristic of seventeenth-century society. Science expressed the values of technological progress, intellectual enlightenment and the glorification of divine wisdom in creating the world. The hostile and mysterious environment of the natural world would yield its secrets to human investigation. The belief in the capacity of human ingenuity to achieve dominion over nature was justified by the argument that the study of God's book of nature was complementary to the study of the Bible, the book of God's word.

These important shifts in cultural outlook dramatically transformed the conception of the cosmos and of man's place in nature. The acceptance of the mechanistic universe and the belief that there might be a plurality of worlds like the earth in infinite space threatened traditional assumptions about the uniqueness of man, leading to a denial of the doctrine that the cosmos had been created for the benefit of man. There was a weakening of traditional beliefs in astrology, witchcraft and magical healing. While the methods of rational science failed to provide effective technical substitutes for magic in the seventeenth century, there were some improvements in human control of the environment. More

2

importantly, there was a growing belief in the potential of man to achieve scientific mastery of the natural world by means of technical innovation. Natural knowledge provided an essential key to human improvement.

Religion played an important role in fostering the belief that an understanding of nature was central to man's destiny. The harmony between the natural and the divine was emphasized, and a commitment to the study of God's book of nature was viewed as complementary to the study of the book of God's word. The understanding of nature and the attainment of technical control over the natural world were regarded as directed towards a religious end. The acquisition of natural knowledge was regarded as a religious duty, the study of divine wisdom as revealed in the universe. The controlled and organized pursuit of scientific knowledge was therefore sanctioned by its ultimate religious ends.

The traditional association between the revealed wisdom of Christianity and the natural knowledge of the philosophers led to considerable intellectual tensions in the seventeenth century. The traditional view of theology as the queen of the sciences expressed the authority of Christianity and its dominance over other forms of knowledge. The radical shift in man's conception of the cosmos in the seventeenth century led to a reappraisal of the authority of theology and the traditional view of its status. While the pursuit of natural knowledge was frequently justified by assertions that Christian doctrine and the study of nature were compatible, and indeed that the rational investigation of God's design of nature was an essential part of Christian piety, the declaration of the independence of natural knowledge from theological control became a commonplace in the seventeenth century. The main thrust of the scientific movement of the seventeenth century was to assert the independence and integrity of the methods and theories of science, while at the same time insisting that the study of nature by human reason would enable man more fully to comprehend God's intentions.

The mechanistic world view of the Scientific Revolution undermined many traditional ideas about man's place in nature. More fundamental than the establishment of any particular theory about the natural world is the change in philosophical perspective which was achieved, a new conception of man's capacity to understand and control the world around him. The idea of man the active operator superseded the notion of man the passive spectator. The scientific movement expressed an essentially optimistic outlook, a belief in the possibility of achieving rational understanding. For the intellectuals of the Enlightenment of the eighteenth century, science provided a model for all rational knowledge. For these reasons the Scientific Revolution is an event of momentous significance in European history.

The medieval world picture

The recovery of Greek learning in western Europe from the Islamic world, which began in the twelfth century, initiated an extraordinary flowering of philosophical debate. By the thirteenth century the universities of Paris, Oxford and Bologna were centres of learning in science and philosophy. The works of Aristotle were studied, and logic, physics, cosmology and mathematics formed the basis of the curriculum. Islamic science and medicine had remained 'Greek' science in that while some improvements in observation, instruments and mathematics were achieved, the essential framework of ideas was determined by the writings of the ancient authors. Aristotle's work was of especial significance in the Islamic tradition, and became so in the west when Greek and Arabic texts were translated into Latin. In the new universities of western Europe the Aristotelian texts and the commentaries upon them formed the basis of study. Many of the questions discussed were those which were raised by Aristotle's physics and cosmology. Nevertheless many of Aristotle's arguments were subjected to a searching criticism by medieval scholars.

According to Aristotle's cosmology the universe was supposed to be spherical with the earth at its centre. The moon, sun and planets were supposed to revolve around the central stationary earth, and were held to be embedded in a layered structure of 'crystalline' spheres. Aristotle emphasized a fundamental distinction in character between these celestial regions and the arena of man's habitation on the central earth. The heavens were unchangeable and perfect and their motion was perfect circular motion. By contrast the earth was the domain of change, generation and decay; here substances were composed of four elements – earth, water, air and fire – each tending to its 'natural place', earth to the centre followed by water, air and fire. This scheme originated in simple observations: stones (supposed as 'earthy') fall while fire rises, and Aristotle maintained that a heavy 'earthy' body will naturally move downward to its 'natural' place, the centre of the universe. Changes in bodies were explained in terms of the causes bringing about change. Thus the cause of a stone falling to earth was its tendency to fall to its natural place; and the stone must be moved by some external cause if thrown through the air. In the writings of Thomas Aquinas (1225–74) Aristotle's physical ideas were used in constructing commentaries on biblical texts, and in attempts to prove the existence of God. Aristotle's theory of causes, that there must be a first cause in nature, was used to support the idea of God who created the universe. Aquinas thus sought to assimilate Christian doctrine to the philosophy of Aristotle.

Theologians were nevertheless aware that many Aristotelian ideas, such as the view that the universe had existed from eternity, contradicted Christian doctrine. This and other doctrines which challenged Christian belief were condemned by the bishop of Paris in 1277, leading to a weakening of confidence in the physical explanations offered by Aristotle. In offering alternatives to Aristotle's theories philosophers did not so much dispute the plausibility of Aristotelian science as question the authority of physical explanations in general. Their objective was to demonstrate that alternatives to Aristotelian doctrine were plausible, and hence to establish the varieties and possibilities of explanation, so as to stress the lack of certainty of all natural knowledge. While Aristotle had argued for the immobility of the earth and the daily rotation of the heavens, Nicole Oresme (d. 1382) maintained that the appearances could be interpreted just as well by supposing the contrary. Oresme did not assert the daily rotation of the earth – indeed he quoted the Bible in support of the doctrine of the immobility of the earth – but offered his argument as a refutation of any attempt to establish reasoned argument and physical theory as valid sources of knowledge. The choice between a stable and a rotating earth must be made on faith rather than by appeal to philosophical argument. Christian doctrine was the only valid authority for knowledge; theology was the queen of the sciences.

Although the general world picture offered by Aristotle remained the accepted theory of the universe, the commentaries of the medieval writers offered plausible alternatives to Aristotle's physics and cosmology, and their arguments were later adopted by Copernicus and Galileo and used with polemical intent. The purpose of these philosophical disputations, which offered alternatives to Aristotle's explanation of the motion of bodies and of the supposed stability of the earth, was therefore to reinforce the authority of Christian doctrine as the ultimate source of knowledge. While there was a realm for physical theory and the study of the natural world, any apparent contradiction between revealed and natural knowledge arose from human misunderstanding. The limitations of human reason implied that Christian doctrine alone could serve as the ultimate authority for human understanding.

Renaissance theories of nature

Historians use the term 'Renaissance' to refer to a complex of events in European history spanning the period from 1300 to 1600, signifying a rebirth in the arts and learning. The difficulty of defining these events and the period itself has given rise to much discussion among historians ever since 1860, when in his *Civilisation of the Renaissance in Italy* Jacob Burckhardt defined the period in terms of the development of the idea of the individual as a personality and citizen, the study of Greek and Roman literary texts, and the discovery of the world and of man through the voyages of exploration, the observation of nature, and the breakdown of medieval ideas about society and nature. The period is generally discussed with reference to art and literary history, but there were important developments in the study of the natural world which echo characteristic Renaissance themes of the rebirth of knowledge, the questioning of Aristotle, the search for the texts of classical antiquity, and the belief in the unity of human knowledge and of man's power over nature.

The interest in natural magic, the control of nature through the manipulation of hidden properties and powers, is a dominating theme in the study of the natural world in the fifteenth and sixteenth centuries. The control and understanding of nature were equated, and magic was conceived as the search for wisdom through the study of divine truths as revealed in the natural world. The natural magic of the Renaissance was an attempt at achieving a harmony between faith and reason. The espousal of the atheistic Aristotle by medieval philosophers was contrasted with the attempt by the Renaissance natural magicians to investigate divine truths in God's book of nature. It was claimed that nature would yield its secrets through the study of its hidden powers aided by divine revelation, not by the study of pagan texts.

The investigation of the natural world by means of alchemy and natural magic was sanctioned by the rediscovery in the Renaissance of a group of texts attributed to Hermes Trismegistus ('thrice-great'). Though in fact written between about AD 100 and 300 and a product of the culture of hellenistic Alexandria, they were believed to be of great antiquity. The 'hermetic' writings contain a remarkable blend of mystic, magical and Christian elements, and were considered to be of ancient Egyptian origin dating from about the time of Moses. These texts were translated c. 1460 by the Florentine scholar Marsilio Ficino (1433–99) and were considered to be the source of Greek philosophy. The best elements of Greek philosophy, especially the ideas of Plato, were viewed as being derived from an ancient source which belonged to the Christian tradition, for the hermetic texts contain many echoes of Christian doctrines. Renaissance thinkers thus secured the harmony between Christian theology and the vision of the natural world presented in the hermetic texts.

The hermetic writings stressed an astrological cosmology in which the terrestrial and celestial realms, the central earth and the spheres of the planets, are connected in a web of affinities and correspondences. Hermeticism postulated an enchanted view of the world, in which matter was held to be impregnated with an active spirit through which celestial influences acted. The aim of natural magic was to grasp the hidden powers of nature and the laws of sympathy and antipathy between material things by activating the planet, metal, gem or plant in which the active spirit was stored. The cosmos was therefore perceived as a unified structure in which the terrestrial and celestial realms were linked by a web of active but hidden powers.

In this pervasive Renaissance conception of nature, which remained influential until c. 1650, man was conceived of as a natural magician endowed with special powers to control the natural world. Man was held to be the focal point of the chain of being between matter and spirit. Man mirrored the universe; there was an analogy between man the microcosm and the macrocosm of the universe. By the activation of spiritual agents by natural magic and alchemy man could control the powers of the universe. The hidden lost wisdom of Hermes would be restored by study of the hermetic texts and by the attempt to trace the hidden affinities and correspondences between the earth and the celestial regions, animals and plants, microcosm and macrocosm. Natural magic sought to exploit the network of hidden powers, revealed for example in the analogy between the brilliance of gems and the brightness of the stars and planets in the celestial realm.

The analogies and correspondences which unified the natural world were generally inferred from simple observations. The fortuitous resemblance between the shape of a root and a human organ would indicate a hidden affinity and suggest the specific use of the root in medicine. Gemstones were regarded as having astrological significance because of their brilliance. The theory of analogies indicated a quite different classificatory system from that employed in modern science, where gemstones, rocks and fossils are distinguished by their physical nature and origin, and not classified by superficial and fortuitous similarities of shape or colour.

Emphasizing that man was the operator of hidden powers in nature, hermeticism also had the implication that mathematics was the key to understanding the hidden essential reality behind visible phenomena. The early Greek philosopher Pythagoras was highly regarded in the Renaissance because of his emphasis on number as the basis of all truth in nature. The belief that the universe was fashioned on numerical principles was fundamental to Plato's philosophy and became an important ingredient in the natural magic current in the Renaissance. The Renaissance belief that fundamental mathematical harmonies determined the structure of nature was crucial in the astronomical work of Copernicus and Kepler.

The hermetic emphasis on the ways in which the powers of the universe could be captured and controlled for human ends was a significant feature of Renaissance natural magic. The study of alchemy became especially important in that alchemical texts claimed to offer secret knowledge about the control of natural processes, while placing an emphasis on evidence gained by observation and the careful cultivation of laboratory procedures. The alchemical texts were closely associated with the hermetic writings, and in the Renaissance the operations of the alchemist were conceived in religious terms, alchemical processes being envisaged as counterparts to Christian rites. The work of Paracelsus (1493–1541) developed alchemical and hermetic ideas, aiming to create a universal science of nature based on the correspondences and analogies uniting the active powers of nature. Man was regarded as uniting in himself the constituents of the natural world, and knowledge was to be achieved through observation and the appeal to divine inspiration. Paracelsus regarded alchemy as providing the key to an understanding of nature. Alchemical concepts offered a means of explaining the composition of matter and provided the basis of a new theory of medicine. The emphasis on observation and the control of nature for human ends, together with Paracelsus' claim that his chemical philosophy would achieve the recovery of knowledge lost at the expulsion of Adam from Eden, were important doctrines which shaped the Scientific Revolution.

Technology and the crafts

Renaissance scholars frequently pointed to three major developments in the mechanical arts: the inventions of printing, gunpowder and the compass, all of which contributed notably to the rapid development of European societies in the period. The invention of the cannon promoted the study of ballistics and a variety of metallurgical improvements, the compass fostered the exploration of the world, leading to an interest in astronomical tables to aid navigation, while the invention of printed books greatly accelerated the diffusion of ideas and learning. In the early seventeenth century men such as Galileo Galilei (1564–1642) and Francis Bacon (1561–1626) stressed the contemporary improvements in the mechanical arts and their relevance to scientific practice. The foundation of Gresham's College in London in the late sixteenth century, where the lecturers emphasized the relations between science and the crafts, encouraged the practical application of mathematical, chemical and medical knowledge. The development of natural magic, especially the work of Paracelsus which exercised an important influence in the late sixteenth century, fostered empiricism, which underlined the crucial role of experience gained through manual operations. The crafts presented new technological information, techniques and apparatus, providing new instruments such as the telescope which, in the hands of Galileo, had a dramatic impact on man's conception of the cosmos. Major technological improvements were achieved in surveying, navigation, metallurgy, dyestuffs and therapeutics, providing a new range of technical problems and techniques.

There is however a great difference between these developments in the crafts and the controlled, systematic, empirical studies of nature characteristic of scientific experiments. Nevertheless the three inventions of printing, gunpowder and the compass provided an important symbol of scientific improvement, indicating the sterility of contemporary academic knowledge and the value to be gained from new sources of knowledge grounded in practical experience. In appealing to the three inventions Bacon urged a synthesis of the empirical and philosophical studies of nature; similarly, in admitting the nobility of the aims of Paracelsus and his followers, he castigated their use of natural magic and appeal to divine inspiration as a perversion of the rational, disciplined, experimental methods that should characterize the study of the natural world.

Witchcraft and popular magic in the Renaissance

The magical cosmology prevalent in the Renaissance dominated both popular and educated culture. There was a widespread belief in hell as a place of physical torment and in the devil as a personality with whom witches made compacts. The rise of Protestantism in the sixteenth century reinforced these beliefs, emphasizing the battle with the devil as an agent of God's judgement. While there was some educated scepticism about the existence of witches, the belief in individuals (generally women) who practised harm by occult means through service to the devil was widespread. Many intellectuals sought to establish the credentials of witchcraft by arguing that witches manipulated the hidden active powers of the universe for evil purposes. The general belief in the power of images and the affinities linking diverse natural bodies served to reinforce the belief of educated men in the potentialities of witchcraft.

Magical beliefs flourished in a society in which human inadequacy was apparent. The precarious food supply, lack of hygiene and sanitation, the limitations in the understanding of disease, all contributed to a society in which there was a low life expectancy. The medical profession was unable to offer effective remedies; treatments such as blood-letting were fundamentally misconceived, and physicians were quite incapable of treating the serious diseases such as typhoid, smallpox and bubonic plague which were endemic. There was a widespread vulnerability to misfortunes such as fire and the failure of crops, and magical beliefs flourished in an environment in which natural catastrophes were frequent. The efficacy of prayer was seen in the popular mind to be akin to magical spells, and the sacraments of the church were associated with magical practices. With the Reformation the magical associations of religious practices were attacked; yet while Protestantism stressed divine providence the village wizard or wise woman offered magic as a means to the improvement of man's lot. Astrology claimed to offer an assessment of the course of human actions by appeal to the motions of the planets, a belief justified by the doctrine of correspondences which supported the notion of the astrological influence of celestial bodies upon terrestrial events. Magical healing by the use of talismans was supported by the views of educated men on the occult powers permeating the natural realm, and on the effect of incantations and charms on physical objects. This enchanted cosmos of the Renaissance was to be dealt a mortal blow by the emergence of the theory of the universe as a mechanical system; together with the gradual increase of human control over the natural environment, this changed intellectual outlook ultimately led to a weakening of traditional beliefs in witchcraft and magical healing.

The conception of nature prevalent in the fifteenth and sixteenth centuries was organic; nature was envisaged as a living organism, of hierarchically structured and interconnected parts. The control of nature depended upon the manipulation of these links. This metaphor of nature as an organism was to be replaced in the seventeenth century by the metaphor of nature as a machine. The revolution in astronomical thought, which shifted the earth from the centre of the cosmos and destroyed the Renaissance view of the astrological cosmos, was of major significance in this development.

The Copernican Revolution

The publication of *On the Revolution of the Celestial Spheres* (1543) by the Polish astronomer Nicolaus Copernicus (1473–1543) initiated a transformation in the conception of the universe. The Copernican theory attempted a restructuring and systematization of astronomy, a comprehensive revision of the mathematical system of Claudius Ptolemy (second century AD) which had formed the basis of astronomical theory until the sixteenth century. Ptolemy's *Almagest* is a detailed and systematic development of the methods of Greek astronomy, based on the supposition of an earth-centred cosmos, with the moon, sun and planets fixed in revolving spheres around the central earth. This picture of the cosmos revealed the structure of the universe but did not in itself account for the complex motions of the planets and moon as revealed even by naked-eye observations. Greek astronomers had therefore devised a complex theory of mathematical astronomy which aimed, by constructing elaborate geometrical models,

12

to provide a precise, quantitative account of the pattern of planetary motions. Ptolemy's *Almagest* represents the culmination of this tradition of mathematical astronomy, and the models and methods employed by Ptolemy shaped subsequent astronomical theory, including that of Copernicus.

The nature of the problem which bedevilled the work of these astronomers can be stated quite simply: the earth is not at the centre of the planetary system, rather the planets (including the earth) follow elliptical orbits around the sun. The full realization of this basic fact of planetary geometry was not appreciated until the work of Kepler in the early seventeenth century. In attempting to chart the motions of the planets in circular paths round the earth Ptolemy was failing to allow for the way in which planetary observations are determined by the motion of the earth as well as by the motions of the planets themselves. Because of the motion of the earth itself around the sun, the paths of the planets as viewed from the earth appear complex and disordered, quite at variance with the assumption of a central, static earth orbited by planets in circular paths.

To account for these observations and yet at the same time maintain the assumption of a central, static earth, Ptolemy employed a complex geometrical model which was designed to represent planetary motions in a manner conforming to the observed data. This geometrical model was not intended to represent the real paths of the planets, but simply to provide a mathematical system, based on complex combinations of circular motions, able to calculate and predict the paths of the planets. Until the time of Copernicus, this attempt to 'save the appearances', that is to predict planetary motions while retaining the Aristotelian theory of a central, static earth, dominated work in astronomy.

The work of Copernicus contradicted the *Almagest* of Ptolemy in several fundamental respects. Copernicus argued that the sun, not the earth, was the centre of the cosmos; that the earth was a planet and, like the other planets, moved in a circular orbit around the sun; and that in addition to its annual motion around the sun the earth rotated daily on its own axis. The assumption of the daily rotation of the earth explained the succession of night and

13

day. The concept of a sun-centred cosmos with the earth in motion immediately explained in principle the way in which the planets appeared to move in complex paths as viewed from the earth. These complex paths were merely apparent, the disorderly nature of the observations was simply the result of the motion of the observer on the moving earth. Nevertheless a problem remained, in that observed planetary motions still did not conform to the Copernican cosmology which supposed that planets followed circular orbits around the sun. The reason is, of course, that planetary orbits are elliptical and not circular. To explain the discrepancies between observations and his sun-centred cosmological theory Copernicus employed many of the complex geometrical devices which had been used by Ptolemy. In the use of traditional geometrical models which explained the observed planetary motions in terms of complex combinations of circular motions, the Copernican theory can be regarded as a restructuring of the *Almagest* on the basis of a sun-centred cosmos.

While Copernican astronomy provides a unified and systematic treatment of planetary motions the Copernican theory did not have any decisive mathematical advantages over Ptolemy's *Almagest*. Copernicus also retained one of the fundamental assumptions of traditional astronomy, the concept of crystalline spheres, though the notion of a sun-centred cosmos was in flagrant contradiction to the prevailing Aristotelian cosmology and physics. The Copernican system is thus a modification of traditional theory rather than a revolutionary break with past astronomical ideas. Nevertheless Copernicus emphasized the contradiction between his sun-centred cosmos and the principles of Aristotelian cosmology. In the first part of his treatise he presented a review of physical arguments which sought to defend the physical implications of the sun-centred cosmology. He drew upon Oresme's arguments in suggesting that the apparent daily rotation of the celestial bodies could be equally well explained if it were supposed that the heavens were stationary and the earth rotated, rather than by supposing the motion of the heavens as viewed from a stationary earth. The physical arguments that Copernicus employed in support of his cosmology were in no sense original or convincing. These arguments had been traditionally employed as plausible conjectures which weakened the authority of Aristotle and hence of the human intellect, and did not carry the connotations of concrete physical theories.

Copernicus' book was seen through the press by a Lutheran clergyman Andreas Osiander (1498–1552), who included his own preface stating that the theory described in the book was to be regarded as a mathematical method to facilitate astronomical calculations. Copernicus would of course have agreed that the complex geometrical devices employed in his book were to be understood merely as a means of calculation, not as describing the real motions of the planets. However, Osiander also implied that the concept of a sun-centred system itself should not be taken as describing physical reality, but was a merely mathematical hypothesis. In this he would seem to be at variance with Copernicus' probable intentions. Copernicus had studied in Italy in the 1490s and there had probably encountered the hermetic view, as expressed by Ficino, of the centrality of the sun in the visible universe as symbolic of God in the whole of creation, an argument which Copernicus himself evokes. Ficino's argument was merely symbolic, while Copernicus presented a sun-centred cosmology, but it is likely that the concept of the central sun as the lamp of the world illuminating and controlling the cosmos expressed Copernicus' basic motivation.

The belief that the centrality of the sun in the cosmos expressed the harmony of nature was crucial to the assimilation of the Copernican system. The leading observational astronomer of the period, the Dane Tycho Brahe (1546–1601), refused to accept the Copernican sun-centred cosmos, though his observations of comets passing through the planetary system called in question the traditional theory of crystalline spheres. In the latter half of the sixteenth century there was therefore some weakening of traditional astronomical ideas, fostered by Brahe's use of new observational equipment. Although Brahe did not adopt a sun-centred cosmology, he bequeathed his observational data to Johannes Kepler (1571–1630), a German astronomer fully committed to the belief that the Copernican system of cosmology revealed the simplicity and harmony of the cosmos. Kepler

15

conceived the structure of the planetary orbits in terms of the perfection of mathematical relationships, seeking analogies between the perfection of geometrical structures and the order of the planetary orbits. He also sought to demonstrate the relationship between the planetary orbits and musical harmonies, seeking to unravel the Pythagorean 'music of the spheres'. This commitment to a cosmology grounded on celestial harmony and geometry illustrates the continued importance of the Renaissance Platonic tradition. Kepler strove for rigorous mathematical accuracy, and his work culminated in the publication of his *New Astronomy* (1609), in which the traditional mathematical astronomy of circular motions was abandoned in favour of the supposition of elliptical planetary orbits. In this new and revolutionary astronomy the Copernican sun-centred cosmos was radically transformed, shaking off the geometry of circles and providing the basis for modern cosmology and physics.

The publication of Copernicus' *On the Revolutions of the Celestial Spheres* initiated a process in which the conception of man's place in nature was transformed. By the end of the seventeenth century the traditional finite, earth-centred cosmos of Aristotle was replaced by a conception of the universe as infinite in extent. Galileo's telescopic observations of the celestial bodies revealed both mountains on the moon and the moons of Jupiter, observations which called in question the Aristotelian doctrine of the fundamental difference in character between the earth and celestial bodies, and revealed that celestial bodies could revolve around a planet (Jupiter) rather than the earth. This bridging of the gap between the earth and the planets contributed to a belief that the earth was not unique, a belief that there was a plurality of worlds like the earth. This growing belief in the plurality of worlds in infinite space threatened traditional assumptions about the uniqueness of man. The earth came to be conceived as a planet and the heavens were robbed of their perfection. The shifting of the earth from the centre of the cosmos, and with it the displacement of man from the focal point of the universe, destroyed the Renaissance conception of correspondences between events on earth and the heavenly bodies. The credibility of the Renaissance astrological cosmos, which depicted an astrological influence of the planets upon human actions, was ultimately destroyed. In 1611 the English poet John Donne lamented that the 'new philosophy' called 'all in doubt'. The new Copernican sun-centred system supposed that the cosmos was 'all in pieces, all coherence gone'. These remarks attest to the difficulty in assimilating the new ideas.

Ultimately the notion of a sun-centred planetary system and the associated ideas (which had not been suggested by Copernicus himself) of an infinite cosmos and the plurality of worlds like the earth in the universe became familiar to educated men. This broader Copernican revolution led to a denial of the traditional idea that the cosmos had been created for the utility of man. In the medieval earth-centred cosmos man was given a central place in the drama of creation, even though the earthly abode of man was the lowliest domain in the universe. The acceptance of the notion of the earth as a planet and of the idea of the plurity of worlds shattered the doctrine of man's unique status in the cosmos. Seventeenth-century writers frequently refer to the humbling of man's pride before the vastness of God's creation, seeing the earth as a mere speck in the amplitude of infinite space. If it were to be denied that the earth alone was the domain of change, generation and corruption, then there could be other worlds similar to our own spread through the universe. Astronomical fantasies became a minor literary genre in the seventeenth century, even Kepler speculating about a voyage to the moon. These cosmic perplexities were sharply drawn by the English astronomer John Wilkins (1614–72) in asserting that the inhabitants of other worlds were redeemed 'by the same means as we are, by the death of Christ'.

Galileo: science and religion

The Copernican sun-centred cosmos aroused little immediate interest from theologians. Copernicus himself was a lay canon of the Catholic Church, and the Church did not react to his work (dedicated to the Pope), which implicitly questioned Aquinas' accommodation between theology and Aristotle's philosophy. There was some interest from Protestant theologians who, like

17

Osiander, emphasized that the new cosmology should be considered as no more than a method of performing astronomical calculations, and affirmed that there was therefore no conflict between Copernican astronomy and certain biblical passages which appeared to imply the stability of the earth in the cosmos. The publications and polemics of the Italian Galileo Galilei (1564–1642) changed this quiescent situation.

Although Galileo had apparently been a convinced Copernican for some years, it was his telescopic observations, described with considerable literary verve in his *Starry Messenger* (1610), that gave a public indication of his astronomical ideas. He described his observations of the surface of the moon, where he noticed changing patterns of light and shade. He observed that these patterns changed in relation to the position of the sun, and he argued by analogy that these spots of light and dark were like the changing shadows cast by mountains on the earth. He concluded that 'the moon is not smooth, uniform, but rough, full of cavities, like the face of the earth'.

Galileo's appeal to an analogy between the earth and the moon had important implications, challenging the prevailing Aristotelian theory which maintained that the earth was the domain of change while the celestial bodies (like the moon) were unchangeable and perfect. Galileo hinted that the existence of mountains on the moon implied a sun-centred cosmology; according to the Copernican theory the earth, moon and planets were all planetary bodies, and if there were mountains on the earth then by analogy (Galileo suggested) there would be mountains on the moon. Galileo's observation of the moons of Jupiter also raised difficulties for the earth-centred cosmology, for the moons revolved around the planet Jupiter rather than the earth, a 'solar' system (as envisaged by Copernicus) in miniature. Although Galileo initially refrained from a direct espousal of Copernicus' theory, his general conclusions were unmistakable.

The *Starry Messenger*, and subsequent astronomical observations by Galileo, caused a sensation, and brought the issue of the Copernican theory to the centre of intellectual debate. Galileo soon fanned the flames of controversy by his barbed polemics, in which before long he explicitly and loudly espoused Copernican cosmology.

Galileo showed little interest in the complex problems of mathematical astronomy that had bedevilled the simplicity which Copernicus had claimed for his cosmology. Indeed Galileo took little notice of Kepler's radical transformation of mathematical astronomy, and continued to espouse a sun-centred system with the planets moving in circular orbits. Galileo's interest was in the basic cosmological pattern of the Copernican sun-centred cosmology and with the problem of creating a new framework of physical principles which would adequately explain the sun-centred cosmos. In place of Aristotle's explanation of the centrality of the earth in the universe as a consequence of a heavy 'earthy' body moving to its natural place, Galileo aimed to create a theory of nature in which physical problems were conceived in mathematical terms. In his later writings on physics Galileo achieved considerable success in this endeavour, demonstrating that the fall of bodies to the earth could be expressed mathematically in terms of a relation between the distance and time of fall, not merely (as with Aristotle) explained in terms of the tendency of bodies to move to their natural place. While Aristotle regarded mathematics as inapplicable to the real physical world, Galileo sought to create a physics which was based on mathematization.

Galileo's disparagement of Aristotle was fundamental to his espousal of the Copernican cosmos, but initially he based his claim to have established the truth of the sun-centred cosmology on his telescopic observations. Galileo's polemics, coming during the period of the Counter-Reformation, soon aroused hostility from Italian Catholic theologians. The attack on Aristotle called in question the Church's acceptance of the broad framework of Aristotle's cosmology as compatible with Christian theology. The claim that the earth was in motion appeared to conflict with some biblical passages which seemed to imply the stability of the earth. Most fundamental of all, in claiming to have established the truth of the sun-centred cosmology Galileo was in conflict with the Church's view of theology as the queen of the sciences, and of the claims of theology to have authority over all other forms of knowledge. The Church emphasized the lack of certainty of all physical knowledge; the choice between competing physical theories could only be made by reference to Christian doctrine. By contrast

Galileo declared that the truth of the Copernican system could be established by science; and that he had indeed demonstrated this fact.

The confrontation between Galileo and the Catholic Church is often represented simply as a conflict between science and religious authority. This interpretation is too simple. The affair also involved an issue of personality. Galileo consistently behaved in an incautious manner, and refused to recognize the difficulty in establishing physical theories as certain truths. Convinced of his omniscience Galileo may well have blundered into a confrontation where a more cautious person might have achieved a compromise. The Church appears to have shown some willingness to accept the sun-centred cosmology as a plausible basis for mathematical calculation, but to have refused to concede that it had the status of an established physical truth. Uninterested in mere mathematical calculation, Galileo insisted that he had established the physical truth of the Copernican theory, and in his *Letter to the Grand Duchess Christina* (1615) he was incautious enough to engage in theological polemics. Galileo argued – by no means in an original, though clearly in an injudicious way – that the language of the Bible was not couched in scientific terms but was written for the comprehension of the ignorant. Thus apparent references to a stable earth did not have the force of physical truths, but merely described the common-sense awareness that the earth appeared fixed, unmoving.

The result was the condemnation by the Holy Office in 1616 of the Copernican doctrines of the sun-centred cosmos and the mobility of the earth as philosophically absurd and heretical in contradicting the Bible. Galileo was formally admonished and Copernicus' *On the Revolutions of the Celestial Spheres* was placed on the Index of prohibited books. Galileo was not completely disheartened and continued to develop his physical arguments in support of the Copernican system. Agreeing to discuss the Copernican system purely as a hypothetical explanation of the planetary motions, he published his *Dialogue on the Two Chief World Systems* (1632), a discussion of the Ptolemaic and Copernican systems of astronomy. The argument of the book was however scarcely impartial but strongly adversarial and contemptuous of both Aristotle and Ptolemy. The admission that the Copernican theory was hypothetical, because God could have ordered the universe in many ways, appeared clearly as merely a sop to Church authority. Galileo was brought to trial by the Inquisition and accused of having disobeyed an injunction of 1616 that he should not hold, teach or defend Copernican astronomy.

It is probable that this injunction, which served as the legal basis for his trial, was a forgery; certainly Galileo himself had a document which authoritatively declared that he had merely been admonished and had not agreed to relinquish any of his opinions. But Galileo had clearly overstepped the agreed basis for discussion of Copernican astronomy. While he was obliged formally to abjure his belief in the Copernican system his books had done much to demonstrate its validity, even though Galileo's commitment to circular planetary orbits began to look old-fashioned with the gradual acceptance of Kepler's theory of elliptical orbits. Whatever the complications engendered by the idiosyncracies of Galileo's personality, the conflict between Galileo and the Catholic Church did highlight an important issue of principle which is characteristic of the Scientific Revolution: Galileo's claim for the independence of natural knowledge from theological control, and his assertion of the integrity and authority of the methods of science. The problem of authority is thus at the heart of the issue. Galileo repudiated the traditional claim that as the queen of the sciences theology had authority over all other forms of knowledge. The methods of science, based on observation and mathematics, could uncover the secrets of nature.

Bacon: the reform of learning

The writings of Francis Bacon (1561–1626), Lord Chancellor under James I, exerted a profound influence on the scientific tradition of the seventeenth century and its cultural framework. Bacon was not himself a scientist but an essayist and moral philosopher. His works such as the *Advancement of Learning* (1605), the *Great Instauration* (1620) and the *New Atlantis* (1627) were both critical of contemporary knowledge and visionary and utopian in offering a scheme for the reformation and reconstruction of learning. The thrust of Bacon's argument was to urge the study of God's book of nature as complementary to the study of the Bible, the book of God's word. The reconstruction of learning would enable science to be directed towards the renovation of mankind, though the methods of the sciences should be stripped of the distortions and encumbrances traditionally exerted by religious authority.

Bacon criticized the learning of the universities where Aristotle was dominant as sterile and arid; the commentators on Aristotle were merely spiders spinning philosophical cobwebs. The craftsmen were like ants, simply collecting information and techniques. Bacon urged that the student of nature should emulate the bee, gathering information and transforming it into a system of knowledge. The method of the sciences would begin with an encyclopaedic survey of all learning, and by the careful collection and comparison of data would sift and classify information about the natural world. The production of knowledge of nature would be based on an experimental method, not simply the collection of data but the careful analysis of information produced by experiment. While 'experiments of light' would reveal general principles, 'experiments of fruit' would ultimately generate practical results which would benefit mankind.

This method would sweep away the impediments to learning which had hampered all previous attempts to understand the natural world. Bacon illustrated these impediments by the metaphor of the four 'idols', false images which distorted our understanding of the real world. The 'idols of the tribe' were the innate features of the human mind, inadequacies of our senses, which led men to error. The 'idols of the market place' were the constraints imposed by the misleading nature of language. The 'idols of the cave' derived from the peculiar education and habits of each individual, which Bacon illustrated by Plato's myth in which a man chained to the wall of a cave can only perceive the flickering shadows of the external world. The 'idols of the theatre' were the false philosophical systems of the past, systems which Bacon claimed would be swept away by his experimental method.

Bacon argued that the new method of knowledge would provide the means to emancipate men's minds from these 'idols'.

22

He castigated the deceits of magic and attacked Paracelsus for attempting to find the truths of nature by appeal to divine inspiration. Magic was a deceit in its pretensions but he recognized that its ends were noble in seeking to control nature. Bacon rejected the obscurity and vainglorious claims of magic: 'man's wits require not the addition of feathers and wings, but of leaden weights'. Science was to be prosecuted in a disciplined and organized manner, placing 'all wits and understandings nearly on a level', by which he meant that in pursuing knowledge, like the bee gathering nectar, man must proceed in a careful and ordered manner.

For Bacon science was to be pursued in a disciplined manner by renouncing the intellectual pride of the philosophers of the universities in appealing to Christian doctrine, however, science was to be directed towards a religious end, a 'restitution and re-investing of man to the sovereignty and power . . . which he had in his first state of creation'. Bacon thus sought the redemption of mankind from the consequences of the Fall of Man; having lost his dominion over nature at the Fall, man must regain it with patient humility through labour and manual operations. Bacon maintained that 'knowledge is a plant of God's own planting'; the restoration of knowledge must be pursued by a 'legitimate, chaste and severe course of inquiry'. The powers of nature would be captured and controlled, as the hermetic texts envisaged, though by a new and appropriate method. Bacon quoted the Book of Daniel in the Old Testament, that 'many shall run to and fro and knowledge shall be increased', in support of his claim that his own age would witness the revival of learning.

Bacon quoted the three inventions of the compass, gunpowder and printing in support of his contention that the achievements of craftsmen pointed to the mechanical arts as the model for collaborative, experimental science. While the aims of the study of nature were essentially religious (the restoration of man's knowledge and dominion over nature in which the spiritual destiny of man was associated with the revival of learning), he rejected the intermingling of natural and religious knowledge characteristic of the false philosophical systems of the past. These aims and methods were to be achieved in the brotherhood of scientists

23

which he termed Solomon's House, where the pursuit of learning was to be cultivated as a means of the renovation of man.

Bacon's importance lies in his vigorous espousal of the experimental method as the key to the understanding of nature, and in locating this method and the gaining of knowledge of nature within the framework of an essentially religious conception of the aims of science. The seventeenth century was a period in which religious attitudes played an important role in sanctioning men's activities. In associating the scientific enterprise with man's ultimate theological obligations Bacon's writings underscored an important motivation that led men to study the natural world. The scientist became the priest of nature.

The Baconian tradition and the Puritan movement

The importance of theological motivations in leading men to study the natural world has led some historians to associate the Calvinist attitude of self-restraint and diligence with an interest in the cultivation of science, and to argue that Puritanism was especially conducive to providing a religious motive for the pursuit of science in seventeenth-century England. This thesis is often employed to explain the importance of science in Protestant Europe and especially in England in the seventeenth century. However, it is apparent that many Puritans questioned the religious value of the study of nature, and that while theological motivations played an important role in fostering the study of nature in early modern England, these motivations (as will be shown below) encompassed a variety of religious allegiances and different approaches to the study of nature. Nevertheless, within this pattern of links between religious attitudes and conceptions of nature, the work of Puritan intellectuals from the 1620s to the Restoration in 1660 is especially important. Appealing to Bacon's writings these men encouraged a commitment to the study of nature as a counterpart to the study of the Bible.

Puritan reformers, notably the Prussian emigré Samuel Hartlib and his circle, stressed the importance of the reformation of education and the study of medicine, technology, agriculture and economic planning, subjects relevant to their belief that the restitution of man's dominion over nature would have profound humanitarian ends. The parliamentary regime during the civil war was sympathetic to these aims and provided some financial encouragement for their achievement. Like Bacon these writers appealed to the prophetic books of the Bible, associating the revival of learning and the return of the dominion of man over nature with the renovation of man. The revival of learning came to be seen as the means for the realization of the utopian paradise, an intellectual regeneration based on experimental science whose ultimate meaning was intelligible in terms of the prophetic texts of the Bible which promised the redemption of man. Their humanitarian, utilitarian programme for the exploitation of scientific knowledge was grounded on the belief in its relevance to social improvement and spiritual enlightenment.

These attitudes, and Puritan affiliations, by no means characterize all aspects of English science in the seventeenth century. Nevertheless the visionary force of Bacon's writings exercised an important influence on the intellectual complexion of the period, attitudes which were fostered by his Puritan followers. The pervasive theological orientation of seventeenth-century English science suggests that these attitudes had an important impact in motivating men towards the study of nature. Many of the prominent English scientists of the period were touched by this tradition; while Bacon's precepts for the pursuit of natural knowledge by the application of the experimental method were frequently lauded, the religious orientation of his programme of scientific enlightenment shaped the continued emphasis on the religious value of natural knowledge.

Descartes: the mechanization of the world picture

The development of the mechanical conception of nature in the seventeenth century constituted the most profound attack on the cosmological assumptions of Renaissance hermeticism. The work of the French philosopher and mathematician René Descartes (1596–1650) provided the most systematic and influential statement of the mechanical view of nature. Descartes conceived his physics as a rational system grounded on philosophical principles whose truth was guaranteed by the veracity of God, a system of science which was conceived as a total refutation both of Aristotelian physics and of the theory of active powers as envisaged by natural magic.

Descartes argued that all of reality is composed of two substances, spirit (a substance characterized by thinking) and matter (a substance characterized by its spatial extension). Spirit possessed none of the properties characteristic of matter, and all material reality was held by Descartes to be inert and devoid of any internal power of activity. All physical substances and phenomena were held to arise from matter in motion, and action by contact between extended portions of matter was the only mode of change in nature. This conception of nature stands in sharp contrast both to the physics of Aristotle and to the theory of matter as permeated by active powers which was characteristic of the natural magic tradition. The rigid separation of matter and spirit, and the emphasis on matter in motion as the basis for scientific explanation, provided a new conceptual framework for scientific theorizing. This mechanistic world view, if not all of Descartes' philosophical principles, had a profound influence on subsequent science, and is fundamental to the Scientific Revolution.

In rejecting the foundations of contemporary learning Descartes sought to subject every idea to rigorous examination by a programme of systematic doubt. He found the basis of certainty in his own existence: 'I think, therefore I am.' From this he concluded that the material world existed, and maintained that the spatial extension of bodies was the basis of all material reality. Properties of bodies such as heat or colour, which Aristotle had attributed to specific qualities of heat or colour in bodies, were not conceived by Descartes as real constituents of the material world. The diverse properties we are conscious of from sense-experience are held to be an illusory guide to the true characteristics of the material world. According to Descartes' mechanical philosophy, bodies are constituted only of particles of matter in motion; apart from extension, the properties we perceive in sensory experience are merely apparent, the result of particles of matter exciting our sensory organs.

In his *Principles of Philosophy* (1644) Descartes attempted to establish that a wide variety of phenomena – magnetism, the structure of the planetary system, and organic phenomena – could be explained in terms of matter in motion. Just as in geometry proofs are deduced from fundamental axioms, so in Descartes' physics mechanical explanations of phenomena are deduced from the first principles of matter and motion. Descartes' theory of matter supposed three 'elements' or fundamental forms of matter distinguished by the different size and motion of their particles, and he explained the rotation of the planets in terms of the flow of smaller particles which carry the planets with them. Based on the concept of matter in motion, this theory provided an alternative physical model of the planetary system to the traditional theory of crystalline spheres. Magnetism had traditionally been regarded as an obvious example of an active power, but Descartes provided a mechanical theory in which screw-shaped particles were held to cause magnetic action. Organic functions were explained in mechanistic terms; animals, lacking a soul, were conceived as mere complex machines.

Descartes' mechanical models were not envisaged as an account of the actual physical mechanisms in nature but merely illustrated the possibility of explaining natural phenomena in terms of matter in motion. Descartes did not seek to discover new phenomena or to investigate nature experimentally, though he did recognize that experiments would be needed to test alternative mechanical explanations. Descartes' intentions were essentially philosophical, to transform the very basis on which physical explanations were constructed. In seeking to refute the assumptions of Aristotelian physics and the world view of natural magic Descartes aimed to establish that nature could be represented by mechanical laws of matter in motion. Descartes sought explanations of the phenomena perceivable in sensory experience by appealing to mechanisms of particles of matter in motion. This became the basic principle of the mechanical world view of the seventeenth century, and while the physical mechanisms which Descartes invented to explain phenomena such as magnetism and the planetary motions were often criticized as being without any empirical support, as being 'fictions', products of the imagination rather than experimental and mathematical demonstration, his importance lies in having established the fundamental basis on which subsequent science was to rest.

The impact of the mechanical philosophy can be seen from an examination of the way in which science was conducted in the period 1650–1700. Mechanical principles were applied to the study of the transmission of light, to the study of chemical reactions in terms of the shapes of chemical particles, and to the study of the human body. The study of the laws of mechanics became central to scientific debate. Mechanical processes, the impact of particles, the operation of levers, seemed to provide the basis for understanding all natural processes.

The Royal Society and the mechanical world view

The foundation of the Royal Society of London at the Restoration of Charles II in 1660 marks an important turning point in the emergence of the new world picture. There was no single scientific orthodoxy which characterized the outlook of its members; indeed attitudes ranged over a wide spectrum from a rigid adherence to Descartes' principles to the continued acceptance of alchemy and natural magic. Moreover the actual activities of the Society were often rather banal. Nevertheless the dominant nucleus of the Society, including Robert Hooke (1635–1703) and Robert Boyle (1627–91), fostered a mechanical approach to the explanation of natural processes based on the application of the Baconian experimental method.

These men owed much to Bacon yet wished to distinguish their view of science as based on the Baconian experimental method from the activities of Bacon's Puritan followers. They proclaimed a vision of the natural world as a mechanical system which could be comprehended by human reason, and maintained that the passions and spiritual frenzies of the dismal age of the Interregnum would be dispelled by the rational study of nature. The study of nature, if properly pursued, would produce belief in the power and wisdom of God. The knowledge of nature would yield a knowledge of God's works, which would lead to a new reverence for God. The reformist zeal of the Puritans, which had tended to the espousal of a view of the natural world based on natural magic rather than mathematics and the mechanical philosophy, would be replaced by the sober, pious, rational and experimental study of nature.

Behind the polemics and apologetics lay some classic investigations in experimental science and interpretations of natural phenomena by appeal to mechanical principles. Hooke's work on optics and microscopy, Boyle's studies of chemistry and the physical properties of gases, were models of careful experimentation. The theory of nature in terms of particles of matter in motion provided the explanatory basis of this work, a model of natural processes amenable to measurement and calculations. Boyle explained barometric pressure in terms of the 'spring' in the particles of air, and by precise experimentation established 'Boyle's law', the relation between the pressure and volume of gases. These works, though scarcely typical of the activities of the Royal Society as a whole, nevertheless exemplified the new style of exact experimentation based on the appeal to the mechanical world view, which became the dominant approach to the study of the natural world in the hands of a new expert scientific élite.

The espousal of the mechanical philosophy did however create intellectual difficulties. In maintaining that the mechanical philosophy aimed to trace all natural phenomena to the two principles of matter and motion, Boyle wished to emphasize that his conception of the mechanical world view was compatible with religious belief. Descartes' rigid demarcation between spirit and matter, and his explanation of nature purely in terms of matter in motion, seemed to some critics to be a 'materialistic' theory incompatible with the Christian view of the universe as subject to divine providence.

To mitigate these doubts Boyle expounded his own version of the mechanical philosophy. According to Boyle nature was constituted of particles, or atoms, which were defined by their shape and hardness. Matter was inert, its activity and laws dependent on the activating agency of the divine will. To explain the relationship between God and nature Boyle employed as a metaphor an analogy between the king and his subjects. Just as the king established laws which subjects obeyed, by analogy matter was dependent for its laws on the providence and will of God.

Boyle defended the explanation of natural phenomena in terms of laws of nature by arguing that the study of nature would reveal the harmony of the cosmos and God's design of nature. Indeed, knowledge of natural laws would enable the scientist, as the priest of nature, to distinguish the truly miraculous acts of God from superstitious wonder at the unusual. Reason and revelation were complementary, one yielding knowledge of the works of God the other knowledge of God's revealed word. Through the mechanical philosophy nature could be understood by reason; indeed, the study of nature by the exercise of reason was envisaged as a religious duty.

Boyle's theological writings indicate the temper of the intellectual tensions of the period. With the development of the mechanical philosophy there had been achieved a new basis for representing nature. The metaphor of the clock, evoked by Boyle and others, provided the new image of the natural world. God became conceived as analogous to a clockmaker; the harmonious workings of nature according to divine laws were compared with the smooth running of a clock. God's power and wisdom were seen as revealed in the design of the clock universe. The mechanistic view of nature and the Christian concept of the deity were thus reconciled.

The Newtonian world view

The mathematical, mechanical physics of Isaac Newton (1642–1727) is generally regarded as the culmination of the Scientific Revolution. In his *Mathematical Principles of Natural Philosophy* (1687) Newton developed earlier discussions of mechanics into his own classic Newtonian 'laws of motion', and provided a coherent cosmology and physics based on Kepler's doctrine that the planets revolve in elliptical orbits round the sun. By appealing to the concept of a gravitational 'force of attraction' acting across empty space Newton was able to provide a secure basis for the new astronomy of the seventeenth century.

Newton was very critical of Descartes' explanation of planetary motion in terms of a fluid permeating the universe and acting on the planets. The complex mechanisms invented by Descartes appeared to Newton to be entirely speculative, without any experimental or mathematical support. Moreover, he was able to show that his own concept of the attractive force of gravity, which was based on a mathematical treatment of planetary motion, was incompatible with Descartes' theory of the solar system. Newton also disliked Descartes' explanation of gravity by means of mere matter in motion. By contrast, Newton claimed that his concept of the force of gravity, rigorously based on mathematical demonstration, could only be explained ultimately as the manifestation of divine agency in nature. The Newtonian cosmology was based on a mathematical mechanics which described the structure of the physical world, but could only be fully understood by means of an appeal to theological principles. Newton claimed that without periodic divine intervention the planetary system would become disordered; nature would run down and needed to be constantly replenished. Following the basic principle of the mechanical philosophy, Newton declared that as matter itself was entirely inert and devoid of all internal activity, the reactivation of the natural world could only be provided by divine agency. Proposing a mathematically precise version of the mechanical philosophy, Newton insisted that his cosmology was fully compatible with the Christian view of the universe.

Newton maintained that the ultimate purpose of the study of nature was to shed light on the problem of God's relation to nature, to understand the cosmos including God as the creator and sustainer of nature. Although he rejected the hermetic view of the cosmos, he supported his theory of gravity by an appeal to the tradition of ancient wisdom, arguing that his concept of gravitation was a restatement of ideas known to Pythagoras which had been hidden in enigmatic phraseology. The task of the student of the natural world was the rediscovery of the true system of the world, the restoration of ancient wisdom by the use of experiment and mathematics. In addition to his physical theorizing and experimentation Newton engaged in biblical and alchemical studies which aimed to penetrate hidden knowledge, to uncover the harmony between natural and revealed knowledge. Newton aimed to achieve a universal knowledge, combining the ancient wisdom, the secrets of the alchemists, the revealed knowledge of

31

ets, and the experimental and mathematical study of
 hile this goal inevitably eluded him, he saw his aim as
ation of a true knowledge of nature so as to restore man
of moral perfection.

    .n conceived the cosmos as unbounded, as infinite in
extent. Here again theological principles were employed to justify
this conception of the universe. Space was conceived as the abode
of the deity; God was present in the infinite space of the Newtonian cosmos and the universe was the 'temple of God'. By
means of the force of gravity God sustained the operations of
nature. The motions of matter were describable by mathematical
laws, but the fundamental workings of the universe were the
result of the activating power of divine agency. The mechanical
universe was like a clock that required rewinding and cleaning.
Newton thus sought to integrate his mechanical philosophy with
his conception of divine providence: natural science alone, being
based on the mathematical analysis of the motions of bodies,
afforded only a partial glimpse into the structure of creation.

Conclusion

The triumph of the mechanistic theory of nature, the clockwork
universe, over Aristotelian physics and cosmology and the tradition of hermetic natural magic, had a decisive influence upon the
intellectual climate. By the end of the seventeenth century there
was a weakening of traditional beliefs in astrology, witchcraft and
magical healing. Educated men were ceasing to regard the
universe as a web of hidden affinities and powers, with man at the
centre dwelling on a unique earth. The credibility of the astrological cosmos was shaken. The organic conception of the universe, which supposed that matter possessed internal active
powers, was displaced by the mechanistic world view, which supposed that matter was inert and devoid of activity. In this new
clockwork universe, experiment and mathematics were the means
to understanding and control.

The Scientific Revolution fundamentally changed man's conception of his place in nature. But if man's pride was humbled in
being removed from the centre of the stage in the drama of

creation, the Scientific Revolution resulted in an unprecedented optimism for human capabilities. While science achieved few significant technological advances before 1850 there was a general belief that scientific knowledge had practical import. In European culture science symbolized rationality, improvement, progress and the promotion of human welfare. Ultimately the theological principles which were so strongly emphasized in the period of the Scientific Revolution itself were detached from the scientific world view: the scientific endeavour came to be seen as theologically neutral, progressive and secular.

Suggestions for further reading

There is a useful introduction to the Scientific Revolution by Herbert Butterfield, *The Origins of Modern Science* (London, 1950). This may be supplemented by two more recent works which present a more detailed account of the period: Allen G. Debus, *Man and Nature in the Renaissance* (Cambridge, 1978), and Richard S. Westfall, *The Construction of Modern Science: Mechanisms and Mechanics* (London, 1971; reprinted Cambridge, 1977). Both these works contains extensive reading lists. On the astronomical revolution the most accessible work is Arthur Koestler, *The Sleepwalkers* (London, 1959), which is however strongly polemical in tone. Michael Hunter, *Science and Society in Restoration England* (Cambridge, 1981), is an accessible study of the place of the scientific world view in the culture of the seventeenth century, while Keith Thomas, *Religion and the Decline of Magic* (London, 1971), provides an exhaustive but highly readable account of the popular culture of early modern England and the impact of the scientific movement upon it. The transition from the organic to the mechanical world view is discussed by Carolyn Merchant, *The Death of Nature* (San Francisco, 1980; London, 1982).

NOTES

NOTES